© 2022 Holly Simon

All rights reserved. No portion of this book may be reproduced in any form without permission from the author, except as permitted by U.S. copyright law.

**ISBN/SKU:** 979-8-218-10129-9

Title: I Came Before the Dinosaurs
Written/Illustrated by: Holly Simon
Edited by: Christopher Flis
Juevinile Ages 6+

# I CAME BEFORE THE DINOSAURS!

### WRITTEN AND ILLUSTRATED BY HOLLY SIMON

My name is Abby, and I am a Dimetrodon. Holly the Paleontologist digs up me and my friends. One thing she doesn't find are Dinosaurs, because I am not a Dinosaur.

I have scaly skin, but that doesn't make me dinosaur. I have sharp claws, but that doesn't make me Dinosaur.

# WHY AM I NOT A DINOSAUR?

## DIMETRODON

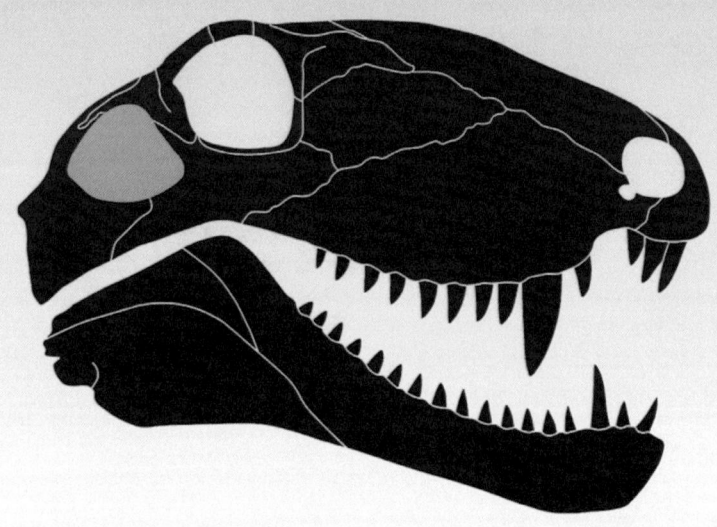

I have one hole behind my eye. Because I have one hole behind my eye, Holly calls me a synapsid.

## TYRANNOSAURUS REX

Dinosaurs have two holes behind their eye. Holly calls them diapsids.

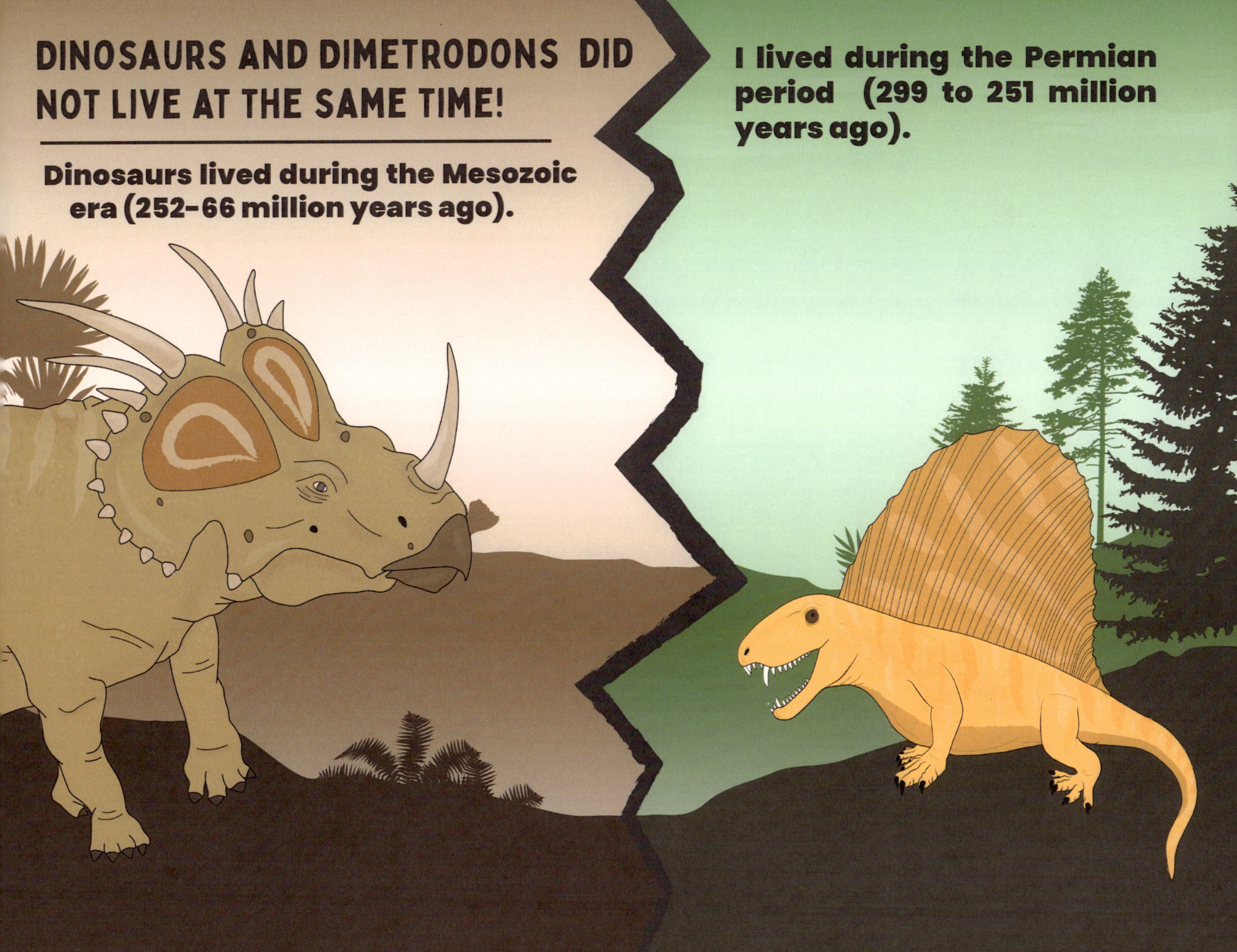

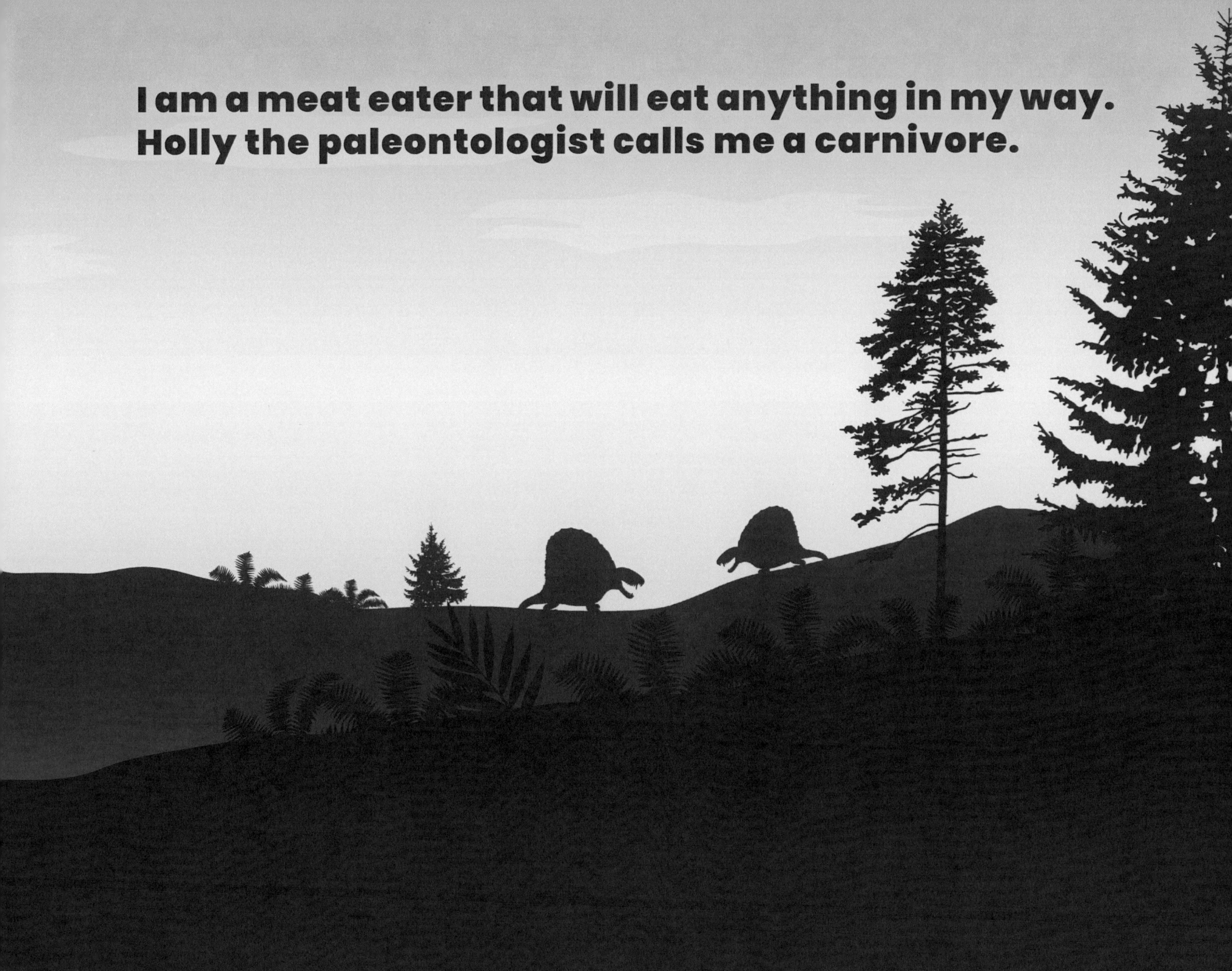

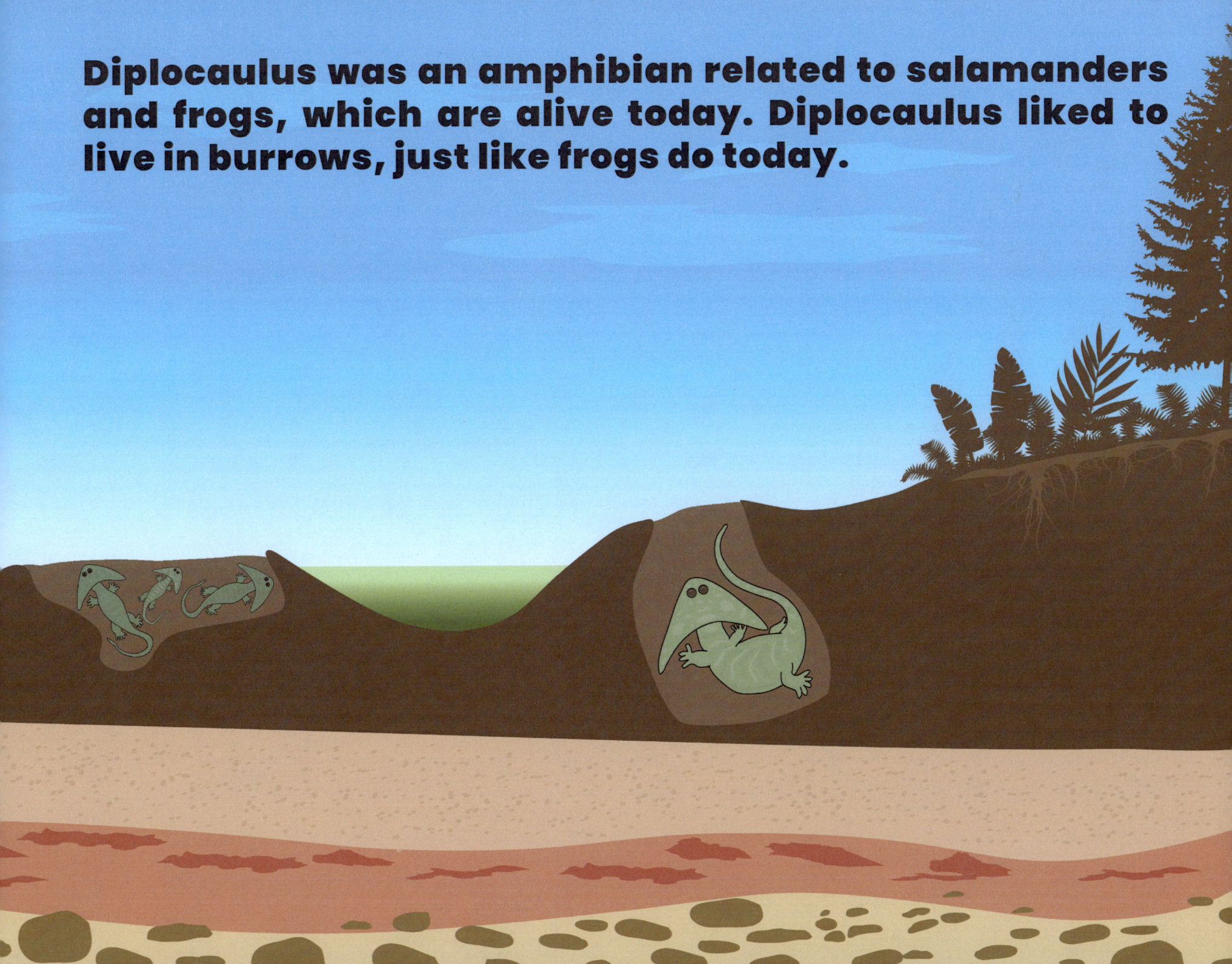

This is Leroy the Edaphosaurus. He eats plants, so he probably doesn't like me very much.

He has a sail on his back like me, but has knobs attached that helped him to blend in with the trees. Probably so he can hide from me.

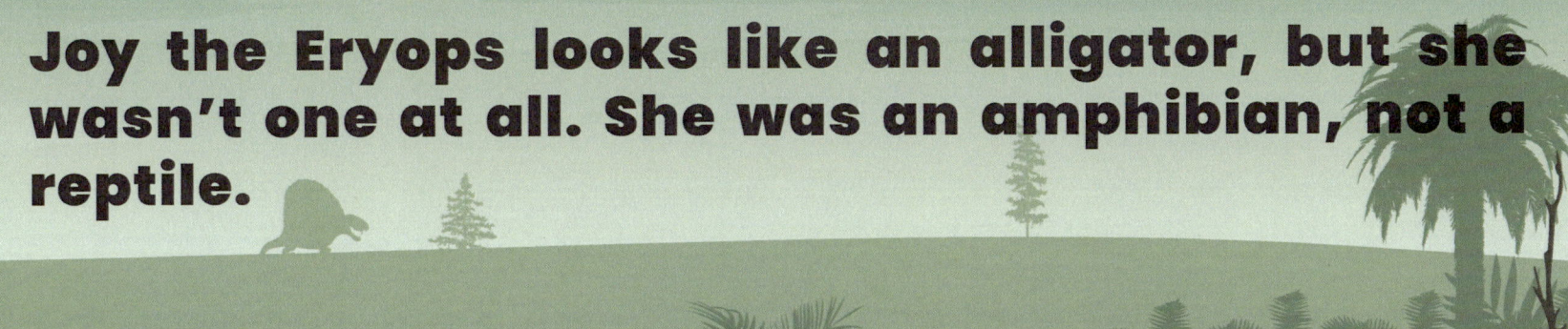

Joy the Eryops looks like an alligator, but she wasn't one at all. She was an amphibian, not a reptile.

**Evelyn is another one of my friends. She looks like me but has shorter legs and a skinnier face.**

Holly calls her a Secodontosaurus. Her name means "cutting-tooth lizard" because she has sharp teeth.

Dimetrodons like me couldn't hear. Plant eaters like Sammie the Diadectes had very good hearing. Probably to hear me coming.

Sammie the Diadectes was an herbivore. She had a large stomach. This is so she could eat lots of veggies.

# DID YOU KNOW THAT NOT ALL SHARKS LIVED IN THE OCEAN?

## Orthacanthus was a fresh water shark that liked to swim in ponds and rivers.

They had a venom-filled spine on the back of their head that helped defend themselves. Probably to protect themselves from me.

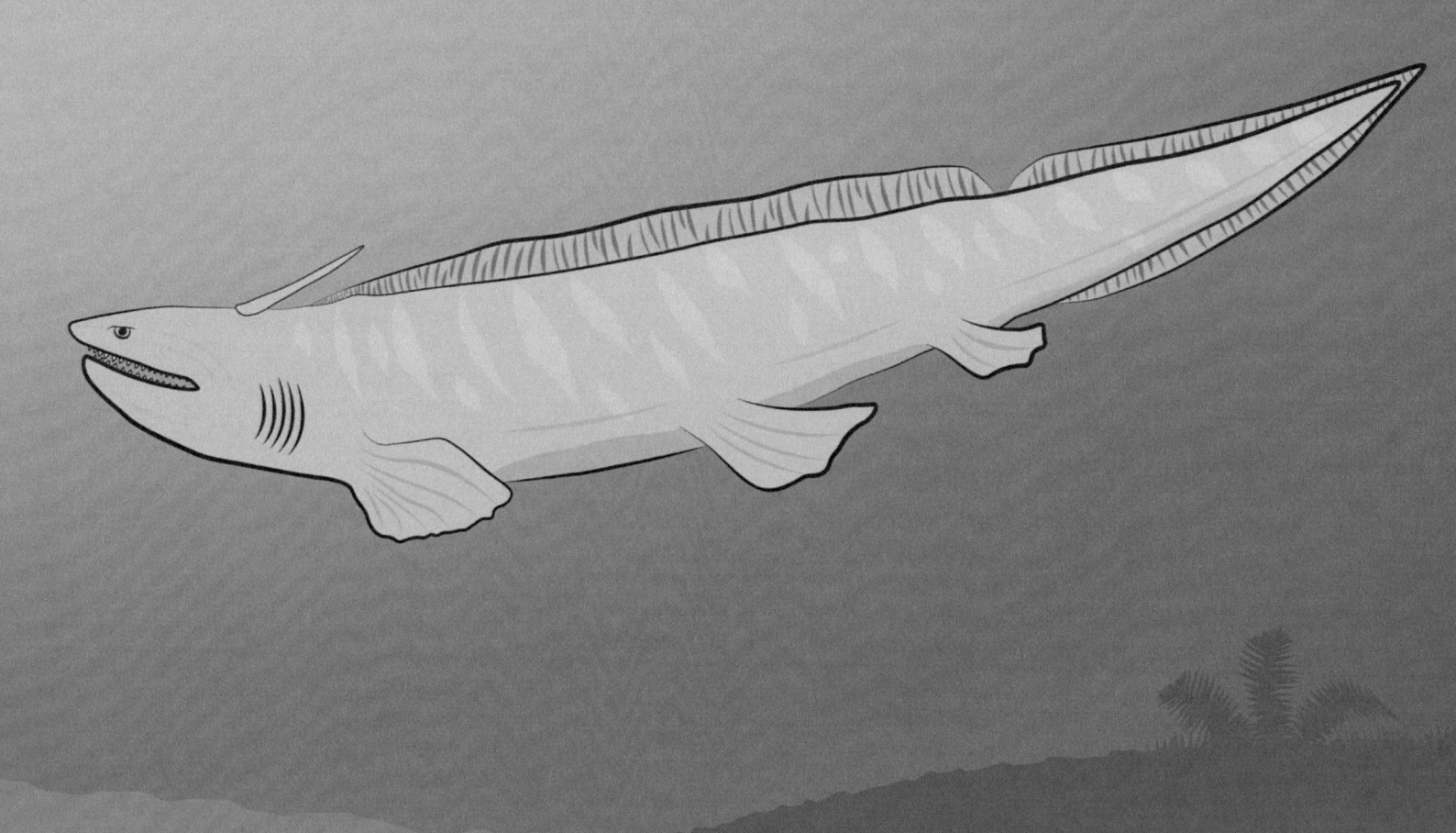

Shelley the Seymouria is special because she is named after the town of Seymour where Holly finds me and all my friends.

Paleontologists like Holly don't know a lot about Seymouria because they are rare. Holly calls her a missing link between reptiles and amphibians.

Now that you have learned about me and all of my friends, you can see us on display at Holly's museum in Seymour.

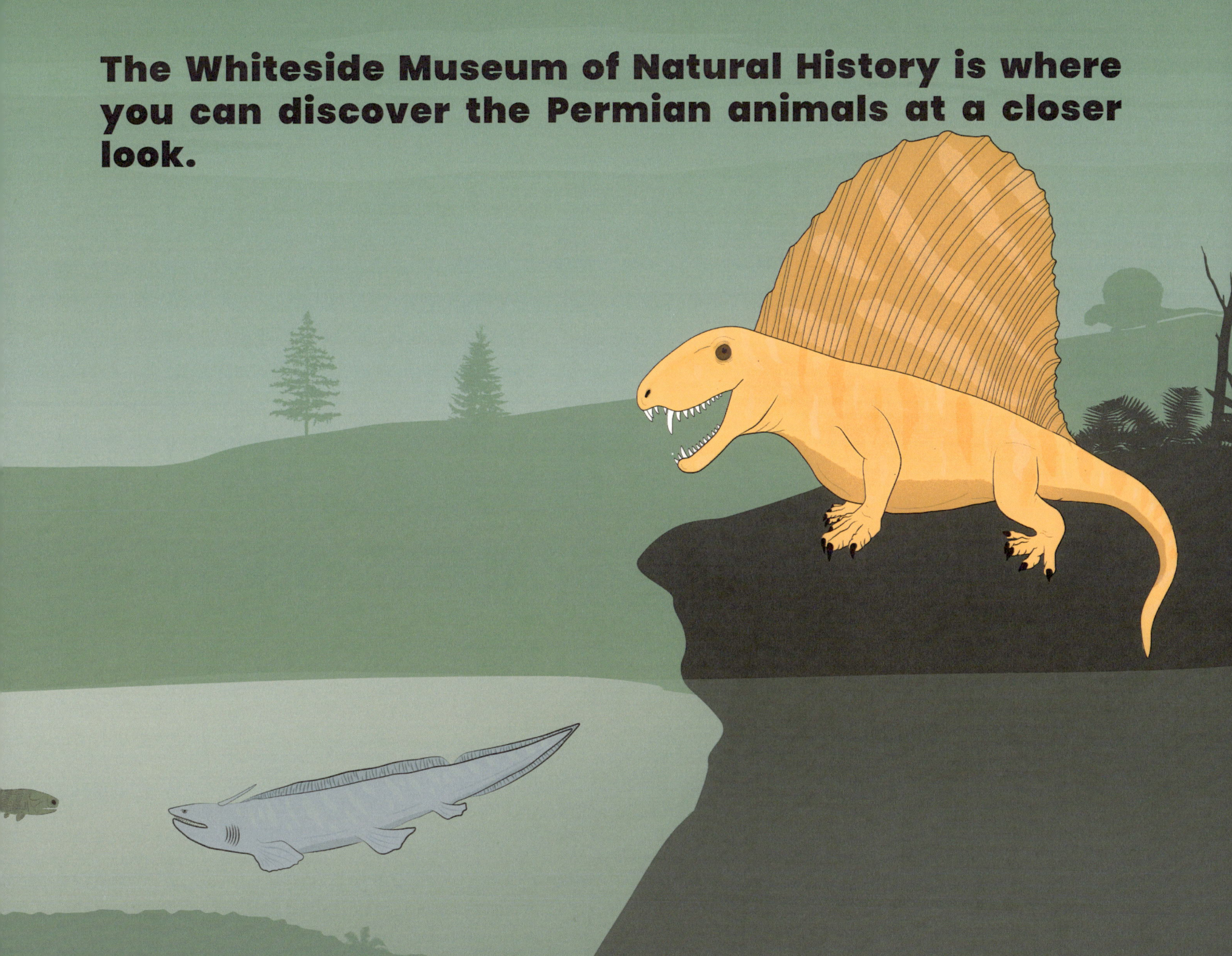

The Whiteside Museum of Natural History is where you can discover the Permian animals at a closer look.

# HOW OLD ARE THE FOSSILS?

**Seymour, Texas is famous for Lower Permian fossils which date back to 287 million years ago. That's about 40 million years before the first DINOSAURS!**

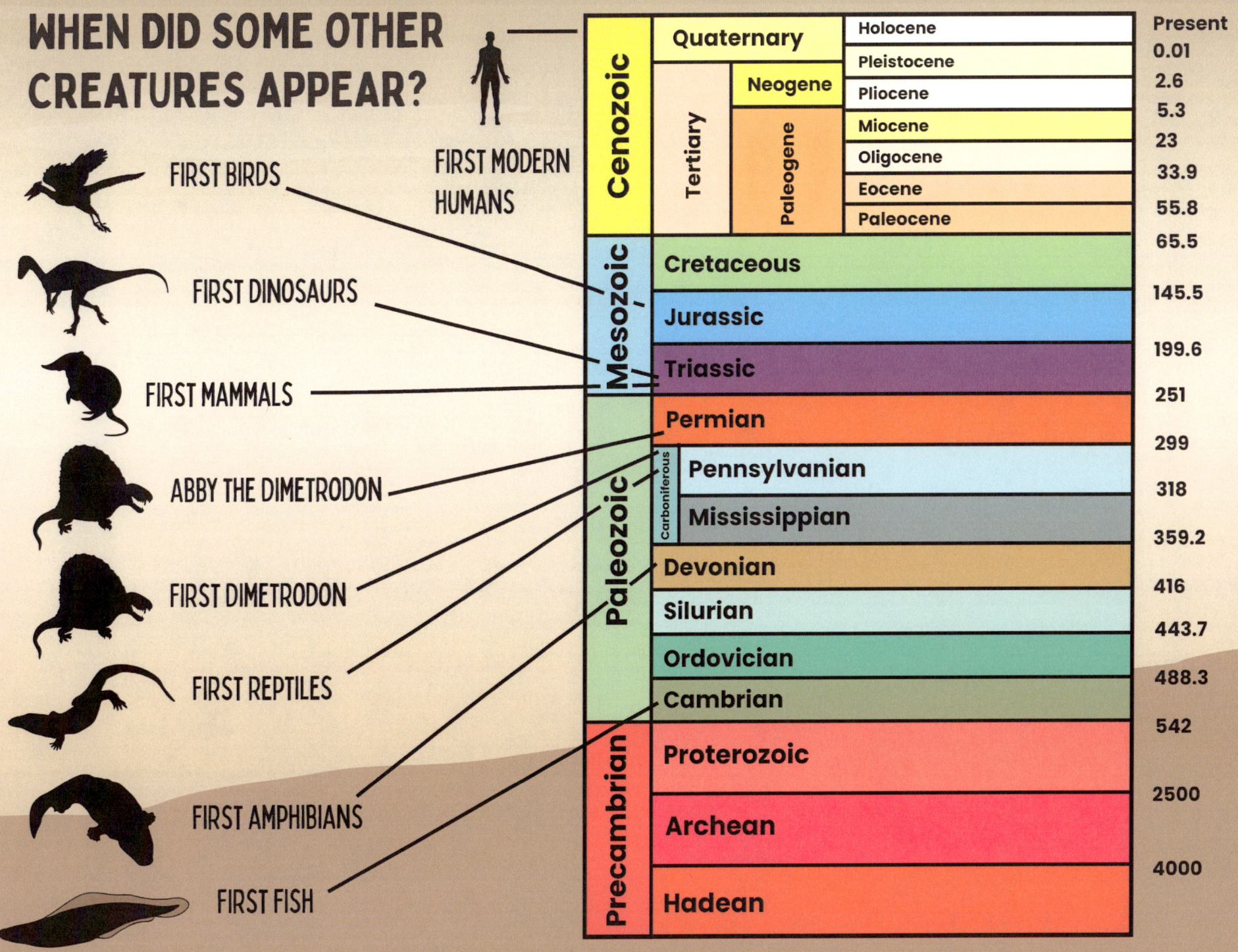

# ABBY THE DIMETRODON

Abby the *Dimetrodon* was discovered by Holly Simon in 2016. Abby's skeleton is disarticulated, meaning the bones are no longer in life position. They have been pulled apart by scavengers after death. The bones with the most muscles are absent, such as the legs and tail where the densest muscles in the body are found. These are the parts that get scavenged first.

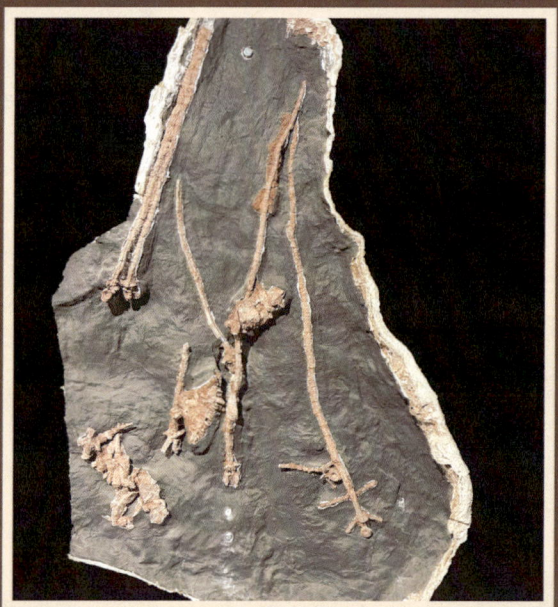

# SANDY THE DIPLOCAULUS

Sandy is one of the many complete *Diplocaulus* skeletons discovered by WMNH paleontologists. This specimen is prepared from the dorsal, or top side of the animal. Chew marks on the tip of the snout and large puncture holes near the eye sockets indicate that this *Diplocaulus* was attacked by most likely *Dimetrodon*. *Diplocaulus* was one of *Dimetrodon*'s main food sources, often pulled from its burrows by hunting carnivores.

# LEROY THE EDAPHOSAURUS

Leroy the *Edaphosaurus* was discovered by Whiteside Museum paleontologists in June, 2015, on the world famous George Ranch fossil beds in Seymour, Texas. On display is the complete spinal column and partial sail of the *Edaphosaurus*. Leroy is one of the largest *Edaphosaurus pogonias* skeletons on record and is also one of the more complete specimens known. Discovered in floodplain sediments, Leroy displays numerous pathologies that indicate he was scavenged by carnivores. As a result, the bones that support the largest muscles are missing. This includes the legs, shoulder, and most importantly, the tail which hosts the largest and densest muscles in the *Edaphosaurus* body.

# JOY THE ERYOPS

Joy the *Eryops* was discovered on June 28th, 2015, by the Whiteside Museum of Natural History less than 5 miles from the museum. The specimen was named after Joy Thompson, a close friend of ranchowners Ken and Charla George. The skeleton possesses one of the most complete *Eryops* skulls ever found in Baylor County. The pointy nose suggests that Joy is male, while the female *Eryops* may have blunt snouts. Joy's skeleton is comprised of a complete skull and lower jaws, plus other elements.

# SAMMIE THE DIADECTES

Sammie the *Diadectes* was found in 2020 by Whiteside Museum of Natural History Assistant Director Holly Simon. Sammie consists of one of the most complete *Diadectes* skulls ever found. *Diadectes* had advanced plant-eating teeth. Unlike carnivore teeth, *Diadectes* teeth were wide and molar-like.

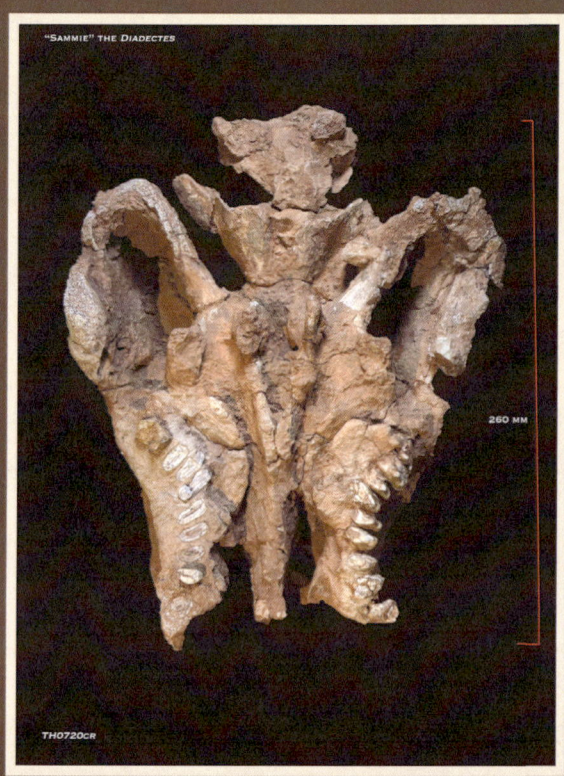

# ORTHACANTHUS

*Orthacanthus* was a species within the Xenacanth shark order that thrived in freshwater ecosystems, such as swamps and ponds, a characteristic much different from ocean dwelling sharks today. This 287 million year old apex predator possessed small, picklefork shaped teeth which supported a fish-eating lifestyle. Its main defense weapon was a long serrated, venom-filled spine protruding from the back of the skull, a fossil that is very common around Seymour, Texas.

# SEYMOURIA

*Seymouria* was first discovered in 1875. It was named after the town of Seymour by Charles Sternberg. While working for Harvard University, Charles Sternberg came across what would be the first fossil evidence of *Seymouria*. The specimen was labeled as "Unknown Reptile No. 1". The specimen was finally officially assigned a name in 1910. *Seymouria* is a significant link between reptiles and amphibians, as it shares traits between the two animal groups. That makes *Seymouria* a little bit amphibian, and a little bit reptile.

# SHELLEY THE SEYMOURIA

*Seymouria baylorensis* is one of the rarest animals in the lower Permian fossil beds of north Texas. Considered a link between amphibians and reptiles, *Seymouria's* skeleton shows similarities to both animal groups. In 2019, museum paleontologist Christopher Flis discovered Shelley the *Seymouria*. It was the first articulated *Seymouria* skeleton found in Seymour, Texas since the early 1900's by Samuel Williston, a professor of paleontology at the University of Chicago.

# WRITTEN AND ILLUSTRATED BY: HOLLY SIMON

WMNH Assistant Director Holly Simon has worked for the Whiteside Museum since 2016 and is both an active member of the paleontological dig crew as well as the museum's Marketing and Social Media Director. Holly has discovered numerous complete *Dimetrodon* skeletons, as well as the most elusive of the Permian animals, *Seymouria*. Holly has also spent many years collecting fossils in the badlands of Wyoming.

## EDITED BY: CHRIS FLIS

Christopher Flis has worked in the museum and paleontology industry since 2004. He was born in Midland, Texas, and has also lived in Dublin, Ireland and London, England. Chris has worked for multiple museums throughout his career, including the Houston Museum of Natural Science, The Black Hills Institute of Geologic Research, and the Lafayette Natural History Museum in Lafayette, Louisiana.

www.ingramcontent.com/pod-product-compliance
Lightning Source LLC
Chambersburg PA
CBRC100812010526
44107CB00023B/1273